Special Report:
Prevention of Self-Contained
Breathing Apparatus Failures

Reported by: Adam K. Thiel

This is Report 088 of the Major Fires Investigation Project conducted by Varley-Campbell and Associates, Inc./TriData Corporation under contract EMW-94-4423 to the United States Fire Administration, Federal Emergency Management Agency.

Department of Homeland Security
United States Fire Administration
National Fire Data Center

U.S. Fire Administration Fire Investigations Program

The U.S. Fire Administration develops reports on selected major fires throughout the country. The fires usually involve multiple deaths or a large loss of property. But the primary criterion for deciding to do a report is whether it will result in significant "lessons learned." In some cases these lessons bring to light new knowledge about fire--the effect of building construction or contents, human behavior in fire, etc. In other cases, the lessons are not new but are serious enough to highlight once again, with yet another fire tragedy report. In some cases, special reports are developed to discuss events, drills, or new technologies which are of interest to the fire service.

The reports are sent to fire magazines and are distributed at National and Regional fire meetings. The International Association of Fire Chiefs assists the USFA in disseminating the findings throughout the fire service. On a continuing basis the reports are available on request from the USFA; announcements of their availability are published widely in fire journals and newsletters.

This body of work provides detailed information on the nature of the fire problem for policymakers who must decide on allocations of resources between fire and other pressing problems, and within the fire service to improve codes and code enforcement, training, public fire education, building technology, and other related areas.

The Fire Administration, which has no regulatory authority, sends an experienced fire investigator into a community after a major incident only after having conferred with the local fire authorities to insure that the assistance and presence of the USFA would be supportive and would in no way interfere with any review of the incident they are themselves conducting. The intent is not to arrive during the event or even immediately after, but rather after the dust settles, so that a complete and objective review of all the important aspects of the incident can be made. Local authorities review the USFA's report while it is in draft. The USFA investigator or team is available to local authorities should they wish to request technical assistance for their own investigation.

For additional copies of this report write to the U.S. Fire Administration, 16825 South Seton Avenue, Emmitsburg, Maryland 21727. The report is available on the Administration's Web site at http://www.usfa.dhs.gov/

U.S. Fire Administration

Mission Statement

As an entity of the Department of Homeland Security, the mission of the USFA is to reduce life and economic losses due to fire and related emergencies, through leadership, advocacy, coordination, and support. We serve the Nation independently, in coordination with other Federal agencies, and in partnership with fire protection and emergency service communities. With a commitment to excellence, we provide public education, training, technology, and data initiatives.

 FEMA

ACKNOWLEDGMENTS

The U.S. Fire Administration greatly appreciates the cooperation received from the following people and organizations during the preparation of this report:

Bruce J. Cavallari	Fire Engineering Magazine
Lou Ann Harris	Scott Aviation
Phil Sallaway	Survivair
Nancy Schwartz	National Fire Protection Association
Charles Smeby	National Fire Protection Association
Richard Stein	Survivair
Bruce Teele	National Fire Protection Association
John Tully	Scott Aviation
Ernie Younkins	International Safety Instruments

TABLE OF CONTENTS

Special Report:
Prevention of Self-Contained
Breathing Apparatus Failures

EXECUTIVE SUMMARY

Self-Contained Breathing Apparatus (SCBA) are one of the most important items of personal protective equipment used by firefighters and rescue personnel. SCBA allow firefighters to enter hazardous environments to perform essential interior operations including offensive fire attack, victim search, rescue and removal, ventilation, and overhaul. They are also used at non-fire incidents involving hazardous materials and confined spaces where there is a threat of toxic fumes or an oxygen-deficient atmosphere.

There have been several well-documented incidents during the past 10 years where SCBA failure may have been a contributing factor in the deaths or injuries of firefighters[1]. These incidents, coupled with a recognition of the importance of self-contained breathing apparatus to firefighter safety, prompted the United States Fire Administration to undertake this study to address any operations trends associated with SCBA failure incidents, and to identify potential problems requiring correction or further study.

Catastrophic failures of SCBA are characterized by the sudden and unexpected failure of any component that would subsequently expose the user to a hazardous environment, or introduce a major complication hindering the ability to escape from the environment. Failures of this nature are relatively uncommon occurrences, especially considering the very large number of route uses of SCBA by firefighters and rescue personnel each day. Although catastrophic failures of SCBA are rare, the evidence suggests that "low-order" failures of SCBA are more common. Examples of low-order failures include freeflowing or improperly connected regulators, improperly tightened or connected hoses, inadequate face-to-facepiece seal resulting in air leakage, or blown O-rings during cylinder changes. These problems are often attributable to operator error or inadequate preventive maintenance. Although these failures may not directly result in firefighter death or injury, they are a concern because they may reduce efficiency or hamper the coordination required for safe and successful operations. For example, a low-order failure may result in delayed ventilation by a truck crew, thus slowing the advance of the engine company which is attacking the fire.

Standards and testing procedures have been changed over time to address problems which led to failures and to ensure that SCBA are more durable and reliable. Nonetheless, firefighters must realize that catastrophic failures of SCBA are still possible. There are limits to the physical and environmental

---

[1] Several of these incidents have been investigated and published as part of the United States Fire Administration's Major Fires Technical Report Series. Of particular note are the reports entitled: *Three Firefighter Fatalities in Training Exercise Milford, MI; Three Firefighters Die in Pittsburgh House Fire;* and *Sodium Explosion Critically Burns Firefighters Newton, MA.*

punishment that SCBA can endure. Regular inspection, upgrade, and preventive maintenance will lessen the potential for catastrophic failures of SCBA. The report identifies a variety of issues and operational aspects of SCBA failures, particularly those related to maintenance and user training. Suggestions for addressing these issues are included throughout the report.

SUMMARY OF KEY ISSUES

Issue	Comments
Failure to Use	One of the most common failures of the SCBA system (i.e., SCBA+Firefighter=System) is the failure to use it. Even with the current emphasis on firefighter health and safety, and the expanding knowledge of the hazards posed by the products of combustion, some firefighters still fail to use SCBA during interior operations in smoke-filled environments, especially during salvage and overhaul.
Hardware Reliability	SCBA that are tested and certified according to the requirements of the NFPA 1981 (1992 and 1197 Editions) Standard on Open-Circuit Self-Contained Breathing Apparatus for Fire Fighters are extremely durable and rugged. Properly used and maintained by well-trained personnel, according to the manufacturer's recommendations, they should provide years of trouble-free service with little potential for hardware failure.
Catastrophic Failures	Catastrophic failures of SCBA resulting in death or injury to firefighters are very rare considering the number of routine uses by firefighters each day. Even if such a failure should occur, the fail-safe design of the SCBA may allow it to function long enough for a firefighter to escape the hazard area.
"Low-order" Failure	Some failures of the SCBA system do not directly result in firefighter death or injury, but may reduce efficiency and hamper fireground operations. This type of failure is relatively common and most often attributable to operator error, physical abuse/neglect, or inadequate preventive maintenance procedures. Examples include: difficult or slow donning of SCBA due to a lack of familiarity or infrequent practice; freeflowing regulators; blown O-rings during cylinder changes; and improperly connected hoses or regulators.
Operator Training	Many low-order failures can be prevented through proper operator training. Before entering hostile environments, firefighters must be appropriately trained in all aspects of SCBA inspection and operation. Continued drilling and practice under realistic conditions must be emphasized until complete familiarity is achieved and maintained. Complete knowledge in the use and limitations of SCBA must become second nature to all firefighters to prevent failures.
Preventive Maintenance	Regularly maintained and tested by competent, properly trained, and certified technicians using the appropriate tools, replacement parts, and testing equipment, following procedures recommended by the respective manufacturers. Fire departments should establish preventive maintenance programs for SCBA to ensure firefighter safety and compliance with applicable regulations. The 1989 edition of the NFPA 1404 Standard for a Fire Department Self-Contained Breathing Apparatus Program can be used to provide guidance for fire department preventive maintenance and training programs.
Upgrades	The 1992 and 1997 editions of NFPA 1981 contain realistic, updated procedures for testing and certification of SCBA. Several changes made in these editions were prompted by failure incidents mentioned in this report. Fire departments should upgrade their existing SCBA to meet the current edition of NFPA 1981, to minimize the risk of repeating past tragedies. SCBA that cannot be upgraded can be replaced with newer models.
Pushing the Edge of the Envelope	Despite the fact that modern, NFPA-compliant SCBA are extremely durable, the materials used in their construction have physical limitations. Firefighters and maintenance personnel must understand that SCBA are not indestructible, and that the potential exists to expose SCBA to factors in the environment that may contribute to or produce failure.

INTRODUCTION

There have been a number of incidents involving SCBA where human errors, component failures, or a combination of factors may have contributed to deaths, injuries, or potential injuries to firefighters and rescue personnel. Many of these incidents have been thoroughly investigated previously, and several are summarized below.

Examples of SCBA Failure Incidents

Milford, MI: Three firefighters were killed when a flashover occurred during a live-burn training evolution in an acquired structure. Subsequent investigation of the firefighters' SCBA provided evidence that rapid component failure (including straps and low-pressure breathing hoses) may have been a contributing factor in preventing their escape from the sudden change in fire conditions. This incident occurred in 1987, before the advent of more stringent performance and direct flame impingement testing procedures for SCBA certification.

Newton, MA: In 1993, 11 firefighters suffered burn injuries in an explosion that occurred while they were trying to extinguish a sodium fire in a metal processing plant. The firefighters were equipped with firefighting SCBA, some of which had older straps made of non-fire resistant materials, and others that had newer, fire resistant harness assemblies. The straps on some of the older units melted from heat exposure and released, while the units equipped with the newer straps continued to function properly. Damage to low-pressure breathing hoses and facepieces was also noted at this fire. Although no current SCBA can provide adequate protection from an explosion of a molten metal like sodium, the newer SCBA performed better in this extreme situation than older units which had not been upgraded to meet the (then current) 1992 edition of the NFPA 1981 standard.

Pittsburgh, PA: Three firefighters died of asphyxiation at a house fire in 1995. All three were wearing SCBA when they entered the house. During the investigation after the fire, the units were tested by the National Institute for Occupational Safety and Health (NIOSH). NIOSH found that while the units were functional and capable of delivering air to the users, they each failed to meet one or more of the performance tests required by NIOSH for certification[2]. Problems were noted with the exhalation values on the units, and also with the calibration of the low-pressure alarms. Only one of four tested SCBA regulators met NIOSH flow rate requirements. (The fourth was from an injured firefighter.) After undergoing the recommended periodic maintenance, inspection, and testing procedures, all four units met or exceeded the functional requirements. These findings help underscore the need for regular inspection and maintenance programs for all SCBA units. Despite these problems, investigators could not determine whether SCBA failure was a primary or contributing factor in the deaths of the firefighters. Since this incident, the Pittsburgh Bureau of Fire has purchased new SCBA and revised its SCBA maintenance and testing program.

Humboldt, CA: In March, 1996, an empty SCBA cylinder was accidentally exposed to a corrosive metal cleaning agent in the bed of a pickup truck while it was being transported back to the fires station from a training exercise. The composite cylinder was refilled and replaced on a fire apparatus. The cylinder failed catastrophically six days later, causing major damage to the apparatus on which it was stored. Fortunately, the firefighters staffing the station were in the dayroom and were uninjured by the blast. This incident underscores the need for training not only in the proper use of SCBA, but

---

[2] USFA Technical Report 078, Three Firefighters Die in Pittsburgh House Fire, Pittsburgh, Pennsylvania.

in proper care and regular inspection procedures. (An article addressing this incident is included in this report as Appendix A, reprinted with permission from the editors of *Fire Engineering* magazine.)

Detroit, MI: An empty SCBA was run-over and caught between the tandem wheels on a ladder truck as it left the scene of a fire, seriously damaging the backplate. After returning to quarters, firefighters refilled the cylinder and placed it on another SCBA mounted in a jumpseat bracket, ready for service. The cylinder ruptured catastrophically several minutes later, causing major damage to the cab of the fire apparatus. If a firefighter had been sitting in the jumpseat, the outcome would have been tragic.

At least two additional incidents, one of which resulted in a firefighter fatality, have occurred recently during air-refilling operations involving breathing air cylinders.

Catastrophic Failures

The catastrophic failure of a SCBA is characterized by the sudden and unexpected failure of any component that would subsequently expose the user to a hazardous environment, or introduce a major complication hindering the ability to escape from the environment. Some examples of catastrophic failures are cylinder rupture, facepiece lens failure, harness failure, and complete regulator failure. Due to the nature of the hazards faced by firefighters and rescue personnel while using SCBA, there is a high probability that the occurrence of such a failure during fire or rescue operations would result in serious injury or death.

Catastrophic failures of SCBA are extremely rare considering the number of units currently in service, and the number of routine uses of these units daily. Most of the recent catastrophic failure incidents were caused by the rupture of breathing air cylinders exposed to an unusual stress like being run over or contacting a corrosive cleaning agent.

It is likely that the continued improvement of testing and certification requirements by the fire service and SCBA manufacturers has resulted in the low number of catastrophic failures of SCBA. Additionally, the incorporation of new materials in the design and construction of firefighting SCBA has improved durability and reliability, further reducing the incidence of catastrophic failures in units that are properly tested and maintained according to manufacturers recommendations.

While catastrophic failures are relatively rare, other types of failures are more common and will be discussed in later sections.

CURRENT TECHNOLOGY

The fire service has benefited greatly from manufacturers' increased use of composite materials and aramid fibers like Kevlar®. This has increased safety and firefighter protection while enabling weight reductions in every element of a firefighter's personal protective ensemble. Nowhere are these benefits more evident than in SCBA. New technologies, fibers, and designs have made SCBA stronger, lighter, and more durable than ever before.

Composite construction techniques and new materials have dramatically decreased the weight of SCBA. Steel breathing air cylinders have been replaced with cylinders constructed from aluminum and reinforced with glass-fiber filaments. Cylinders are reinforced with Kevlar® and other composite materials to increase strength while further reducing the weight of SCBA. The recent introduction of carbon-fiber breathing air cylinders for SCBA will allow further weight reductions. This improves safety by enabling firefighters to perform more work before requiring rest and rehabilitation. The ability to withstand higher pressures has made available low-profile cylinders that increase firefighters ability to maneuver and help reduce the possibility of entanglement or entrapment. Higher pressures also translate into SCBA with cylinders having up to one-hour durations. SCBA that incorporate these improvements have been available for several years.

The continued emphasis on firefighter health and safety has led to improvements in facepiece design, including a choice of different sizes to properly fit differently shaped faces), features to improve voice communications, and even built-in units that display information directly on the facepiece such as the amount of air remaining and the positive-pressure status. Facepiece fit-testing has become a critical component of fire department SCBA programs, and should be conducted annually, according to the requirements of the NFPA 1404 (1989 Edition) *Standard for a Fire Department Self-Contained Breathing Apparatus Program*.

As a result of several incidents in which harness assemblies failed due to intense heat and/or flash-over conditions (e.g., those previously mentioned in Milford, MI and Newton, MA), SCBA straps are no longer constructed from materials such as nylon or polyester which have low heat tolerance. Newer straps are fire resistant and most incorporate heat-tolerant aramid fibers in their construction. Some strap designs are further reinforced with steel cable to ensure that the SCBA stays in place during extreme conditions. Fire departments should upgrade their straps to reduce the possibility of harness assembly failure during firefighting operations.

Positive-pressure regulator assemblies have become more reliable and easier to maintain with improved designs and the increased use of lightweight alloys and spaceage polymers. These newer units are more durable, resistant to failure in extreme fire conditions, and less subject to malfunction from adverse environmental factors such as extreme cold or heat. Many of these regulators have fewer moving parts and are easier and less costly to maintain than older units. **Demand-type regulators are still in use, despite the fact that they do not provide adequate respiratory protection to firefighters and rescue personnel operating in hazardous environments. Fire departments that are still using demand-type SCBA should replace them as soon as possible with positive-pressure type units. The use of demand-type regulators for structural firefighting is prohibited by 29 CFR 1910.156(f)(2) (the OSHA Fire Brigade Standard) and is contrary to the NFPA 1404 (*Standard for a Fire Department Self-Contained Breathing Apparatus Program*) and NFPA 1500 (*Standard on Fire Department Occupational Safety and Health Program*) standards.**

SCBA manufacturers offer a range of options including integrated PASS devices and "Buddy-Breathing" systems. The desirability and use of these options is an individual decision for each fire department and a subject of some debate within the fire service. Nonetheless, the widespread availability of these options is a positive sign that manufacturers are taking a renewed interest in firefighter safety by responding to customer requests with innovative improvements to SCBA.

When properly maintained, tested, and used by well-trained firefighters, modern SCBA that comply with the NFPA 1981 (1997 Edition) *Standard on Open-Circuit Self-Contained Breathing Apparatus for Fire Fighters* should provide years of trouble-free service. Although SCBA have greatly improved during the past 10 years, future advancements are on the horizon that should help to further enhance the safety of firefighters and rescue personnel.

Unfortunately, not all fire departments have been able to keep pace with the revised standards, and as a result are still using out-of-date, less functional, or improperly maintained SCBA. These SCBA should be upgraded or replaced to avoid the possibility of repeating past tragedies.

REGULATIONS, STANDARDS, AND TESTING

There are a variety of regulations and standards pertaining to the respiratory protection of workers in hazardous environments, and specifically to fire departments.

Federal Regulations

There are several pertinent Federal regulations affecting SCBA. The enforcement of these regulations has been entrusted to several different agencies, including the National Institute for Occupational Safety and Health (NIOSH), the Occupational Safety and Health Administration (OSHA), and the Mine Safety and Health Administration (MSHA). These regulations are general in nature and apply to all types of respiratory protective equipment, including SCBA.

National Institute for Occupational Safety and Health (42 CFR Part 84)--On July 10, 1995, NIOSH made effective a "Final Rule" on respiratory protective devices in which NIOSH assumed sole responsibility for the testing and certification of most types of respiratory protection equipment for industry, including fire department SCBA. The performance-based testing and certification procedures incorporated in 42 CFR Part 84 replaced less stringent methods used previously.

Mine Safety and Health Administration--Until 42 CFR Part 84 became effective in July, 1995, MSHA had joint responsibility with NIOSH for the testing and certification of all types of respiratory protective devices, including fire department SCBA. With the publication of the "Final Rule", MSHA is now responsible only for respiratory protection equipment intended for use in mining. NIOSH and MSHA will continue to jointly certify respiratory protective devices intended to be used in mining applications.

Occupational Safety and Health Administration--On January 8, 1998, OSHA promulgated revised regulations on respiratory protection equipment and programs. These regulations, which are contained in 29 CFR 1910.134, apply to workers in private industry, Federal government employees, and employees of State and local governments in the 25 States which operate OSHA-approved occupational safety and health programs. The applicability of these regulations to volunteer emergency responders has not yet been fully addressed in all States.

These revised regulations contain specific provisions for fire departments with respect to respirator usage during interior structural firefighting, including the requirement that, "all personnel engaged in interior structural firefighting use SCBA respirators."[3] Furthermore, 29 CFR 1910.134 states "that workers engaged in structural firefighting work in a buddy system: at least two workers must enter the building together, so that they can monitor each other's whereabouts as well as the work environment."[4] OSHA also requires that, during interior structural firefighting operations, "there be at least two standby personnel outside the respirator use area, i.e., outside the fire area."[5]

29 CFR 1910.134 addresses, "elements of respirator maintenance and care that OSHA believes are essential to proper functioning of respirators for the continuing protection of employees."[6] Included

---

[3] Federal Register: January 8, 1998 (Rules and Regulations), pp. 1245

[4] ibid.

[5] ibid

[6] ibid., pp. 1248

among these are cleaning and disinfecting, storage, inspection, and repair. Appendix B-2 of 29 CFR 1910.134 provides specific procedures to ensure that employers, "provide each respirator wearer with a respirator that is clean, sanitary, and in good working order."[7] (It should be noted that, according to this OSHA regulation, cleaning and disinfection procedures provided by the manufacturer may also be used, provided they are as effective as those detailed in Appendix B-2.)

SCBA Standards

Several industry standards have been created in addition to the Federal regulations summarized above. The American National Standards Institute (ANSI) and the National Fire Protection Association (NFPA) each have developed standards. Respiratory protection issues in general are addressed by ANSI and standards for fire department use have been developed by NFPA. The ANSI standards are voluntary, consensus-based standards that are widely used by business and industry, as are the NFPA standards which are specific to fire departments.

The NFPA standards are voluntary, consensus-based standards relating directly to fire department Self-Contained Breathing Apparatus and SCBA programs. Fire departments that adopt the NFPA standards are more likely to meet State or Federal regulatory requirements. Several key standards are summarized below.

ANSI Z88.2-1992: *Respiratory Protection*--This ANSI standard is primarily intended for industrial users of respiratory protection equipment including air-purifying particulate respirators (APRs), powered air-purifying particulate respirators (PAPRs), other types of filter masks, cartridge respirators, and SCBA.

NFPA 1981 (1997 Edition): *Standard on Open-Circuit Self-Contained Breathing Apparatus for Fire Fighters*--This standard, which is approved by ANSI, contains additional requirements beyond the certification requirements mandated by NIOSH. This standard is specific to open-circuit type SCBA for firefighting purposes like those most commonly employed by fire departments. The standard details the technical requirements and testing procedures that must be followed for certification under the NFPA standard.

The testing requirements in the 1997 and 1992 editions of NFPA 1981 were made more rigorous based on evidence that SCBA certified under previous editions were subject to failure under extreme conditions including flashover. As a result, the new testing procedures are more reality-based, and should help further improve SCBA safety, reliability, and durability.

There are two features of the recently released 1997 edition of NFPA 1981 that deserve special mention. First is the requirement that SCBA manufacturers receive ISO 9002 certification for their manufacturing processes (a requirement that many manufacturers already meet). Secondly, SCBA receiving certification under the 1997 edition of NFPA 1981 must be equipped with at least two different types of low-air alarms. The two alarms must function independently (i.e., the failure of one alarm will not affect operation of the other) and must address different senses (i.e., an audible alarm could be accompanied by a visual or tactile alarm).

NFPA 1404 (1989 Edition): *Standard for a Fire Department Self-Contained Breathing Apparatus Program*--This standard, created by the NFPA Fire Service Training Committee, details the requirements for a comprehensive fire department SCBA program including training and preventive maintenance pro-

---

[7] ibid

cedures. This standard is also approved by ANSI, and meets or exceeds the requirements of NIOSH and OSHA regulations.

NFPA 1500 (1992 Edition): *Standard on Fire Department Occupational Safety and Health Program*--This standard, prepared by the NFPA Technical Committee on Fire Service Occupational Safety and Health, addresses a variety of issues related to protective clothing and equipment. Chapter 5-3 of NFPA 1500 relates specifically to SCBA and contains requirements for fire department SCBA programs.

SCBA Testing and Certification

To ensure compliance with applicable Federal regulations, and to receive certification under the 1997 edition of the NFPA 1981 standard, self-contained breathing apparatus are subjected to a battery of tests. The testing criteria included in the current edition of NFPA 1981 are designed to be more realistic than past tests. As a result, SCBA receiving third-party certification under the 1997 and 1992 editions perform at a higher level than ever before required.

Several of the incidents mentioned in this report helped form the impetus for these more rigorous testing and certification requirements. The incidents in Milford, MI and Newton, MA, along with several other incidents that did not result in firefighter death or serious injury, proved that SCBA tested under previous editions did not perform adequately under extreme conditions like those found in a flashover.

The 1997 and 1992 editions of NFPA 1981 greatly improve upon previous SCBA testing methods. Prior to these editions of the standard, each component was tested individually, and there was little provision for simulating realistic fire conditions during the tests. As a result of several incidents involving the failure of specific components (e.g., straps, facepieces) when subjected to high radiant heat or flashover conditions, the standard now demands that all SCBA undergo a direct flame contact test. SCBA certified under the NFPA standard must continue to function while exposed to direct flame contact at a temperature of at least 1,742 degrees Fahrenheit for 10 seconds.

Other tests that were changed include those measuring the ease of voice communication through the facepiece, and the lens abrasion resistance test. These tests were modified to be more objective, realistic, and reproducible.

One of the most important components of the 1997 and 1992 editions of NFPA 1981 is the requirement that SCBA obtain third-party certification from a recognized, independent party (e.g., the Safety Equipment Institute or Underwriter's Laboratories). Previously, manufacturers self-certified their product. This requirement ensures that all SCBA receiving certification undergo identical tests. This is accomplished by having the unit tested at an independent laboratory, or by having a representative of the certifying party at the manufacturer's testing facility for each test.

The testing and certification requirements in the 1997 and 1992 editions of NFPA 1981 exceed the regulatory requirements imposed by NIOSH under 42 CFR Part 84.

Breathing Air Cylinder Testing

The testing of compressed gas cylinders, including those containing breathing air for SCBA, is regulated by the United States Department of Transportation (DOT) under CFR 49, subpart 173.34 (e).

Breathing air cylinders must be hydrostatically tested on a regular basis. Individual maintenance and test records for each breathing air cylinder should be kept to assist departments with meeting the required test intervals. Test intervals for SCBA cylinders are dependent on construction type. Composite SCBA cylinders must be tested at a minimum of every 3 years. Cylinders constructed from aluminum or steel must be tested every 5 years. All composite SCBA cylinders have a maximum service life of 15 years, provided they are hydrostatically re-tested on a regular basis in accordance with DOT regulations. At the end of the 15 year service life, composite cylinders must be removed from service and destroyed (to ensure that they cannot be used again). While aluminum or steel cylinders have an indefinite service life they must be carefully monitored for signs of damage, especially around vulnerable areas like the cylinder neck.

The above are minimum requirements. Cylinders have been subjected to atypical stresses (e.g., dropped from a height, run-over by fire apparatus, etc.), or exposed to chemicals such as corrosives must be immediately removed from service, purged of air, and hydrostatically tested. Failure to ensure that cylinders are properly hydrostatically tested as prescribed in DOT regulations may result in catastrophic failures that could kill or injure firefighters, as well as cause major damage to apparatus and facilities. The incidents in Detroit, MI and Humboldt, CA illustrate the importance of training personnel to recognize cylinders requiring hydrostatic testing.

Personnel tasked with filling SCBA cylinders must be aware of hydrostatic testing requirements. All cylinders with a current hydrostatic test must have the date of the last test either stamped on the cylinder neck, or prominently displayed on the body of the cylinder. Before filling a breathing air cylinder, it should be visually inspected for damage and checked for a current hydrostatic test date. Cylinders that have visible thermal or mechanical damage, or that lack a current hydrostatic test should never be filled. These cylinders should be purged and immediately removed from service. The cylinders may then be sent for hydrostatic testing by qualified personnel.

Although it seems a routine activity, extreme caution must always be exercised during air refill operations. Personnel tasked with refilling SCBA cylinders should be properly trained and educated about the potential dangers of compressed air. In addition, personnel should maintain awareness of any safety advisories or bulletins issued by NIOSH that pertain to SCBA. (These advisories are often mentioned in trade publications like Fire Engineering.) Strict procedures should be followed to ensure that fill-lines are not charged inadvertently, and that cylinders are filled to the correct pressure. Only Grade D breathing air with a low moisture content should be used to fill breathing air cylinders. Air samples should be regularly tested to ensure quality and prevent impurities from contaminating breathing air supplies. Filters should be changed at intervals suggested by the manufacturer, or more frequently depending on the results of air quality testing. Despite the safeguards built into most air refill systems, accidents can occur from improper training or carelessness.

An article in the October 1996 Fire Engineering magazine (included in this report as Appendix B, reprinted with the permission of Bruce J. Cavallari and Fire Engineering) answered many commonly asked questions regarding the hydrostatic testing and certification of compressed gas cylinders[8].

PREVENTIVE MAINTENANCE

Many of the complaints received by NIOSH about SCBA problems are directly attributable to a lack of, or improperly performed, preventive maintenance. Although modern, NFPA-compliant SCBA are both durable and reliable, they require regular inspection and maintenance to maintain these characteristics.

[8] Cavallari, Bruce J. "Hydrostatic Testing of Compressed Gas Cylinders." Fire Engineering, October 1996: 80-82

Every manufacturer has recommended preventive maintenance procedures for their SCBA. These procedures are included with the unit when it is shipped, and many manufacturers will also have a field service representative personally review the procedures with the fire department. The manufacturers' maintenance recommendations should be considered a minimum. SCBA that are subjected to frequent use may require maintenance at more frequent intervals. Most manufacturers recommend annual flowtests. At least one manufacturer adds the additional recommendation of rebuiliding the unit every 3-6 years, or more often if subjected to frequent use.

Preventive maintenance procedures should only be performed by properly trained and qualified personnel who have been certified by the SCBA manufacturer as capable of safely performing such work. Some fire departments have a dedicated "Air Shop" where SCBA are serviced by certified technicians who are fire department employees. Other departments may send SCBA to a private firm that specializes in servicing their particular units. Unfortunately, many departments have not adopted the recommended procedures regarding routine SCBA maintenance. Routine preventive maintenance is the only way to ensure that SCBA will properly protect firefighters.

Another problem is posed by the so-called "midnight mechanics" who attempt to repair SCBA without the proper training, tools, or replacement parts. Although SCBA are becoming easier to maintain, they are still complex apparatus whose reliability depends upon maintenance performed by skilled technicians. The legal ramifications of allowing a non-certified person to repair SCBA could be severe should a failure occur. Knowing that all of the fire department's SCBA units are properly maintained and inspected by competent personnel can also enhance the confidence and safety of the firefighter or rescuer who must face extreme conditions while wearing the SCBA.

Routine SCBA maintenance tasks such as daily inspection, cleaning, and disinfection are usually handled by field personnel, and are at integral part of overall preventive maintenance. However, care must be taken to ensure that they are properly performed. The use of cleaning solutions or procedures other than those recommended by the manufacturer could result in damage to the unit.

To prevent problems from occurring during daily inspection and cleaning, firefighters should be trained on the proper cleaning procedures and solutions to use. Firefighters should also be trained to recognize the signs and symptoms of problems requiring that SCBA be removed from service and referred to qualified service technicians. Examples of these may include: "sticky" inhalation or exhalation valves; leaking, cracked, or heavily abraded facepieces; frayed straps or broken buckles; and any signs of physical or chemical damage to the harness or cylinder. Although something like a frayed strap may not seem severe during a daily check, it may cause a problem for a firefighter during donning or doffing, especially in an emergency situation.

Some common problems with SCBA are related to their storage. SCBA should be kept as clean as possible, which may preclude storage on parts of fire apparatus that are exposed to weather and road dirt or grime. Facepieces should be stored in bags to ensure that they remain clean and ready to use. Particular attention should be paid to ensuring that connections are properly tightened, since road vibration may loosen vital connections over time.

Fire departments that do not have a comprehensive SCBA program, including preventive maintenance procedures, in place can become familiar with NFPA 1404, *Standard for a Fire Department Self-Contained Breathing Apparatus Program*. Using this document as a blueprint will help departments implement their own program to ensure compliance with regulations. More importantly, a well-designed and implemented SCBA program will help ensure that firefighters are venturing into harm's way with safe and effective protective equipment.

TRAINING

As with any area of fire department operations, there is no substitute for proper training with SCBA. Firefighters need to be familiar with SCBA operation while performing tasks like fire attack, victim search, rescue, ventilation, salvage, and overhaul. Firefighters and rescue personnel should also be trained in routine preventive maintenance procedures like daily inspection, cleaning, disinfection, and cylinder changes.

Unfortunately, the tendency exists to treat SCBA as a tool more akin to an axe or pike pole than a vital piece of personal protective equipment. Even in busy fire departments, firefighters are likely to perform only a cursory check of their assigned SCBA when going on duty. In order to prevent "low-order" failures from occurring, it is imperative that firefighters thoroughly inspect and test the function of SCBA as often as possible. In a career department this should happen every day, and should be second-nature. A volunteer department should make SCBA inspection and practice part of every scheduled meeting, drill session, or training class.

SCBA training is essential for all firefighters. Basic familiarization and orientation to SCBA should be part of every firefighter's initial indoctrination. Members should receive extensive training in the use of SCBA under low-stress conditions, as well as during simulated fireground operations. This training should be conducted according to recognized standards, using methodology approved by the appropriate fire training agency. Individual training records kept for each firefighter or SCBA user will help ensure that members are maintaining proficiency with SCBA.

Basic training in SCBA use should include a discussion of the limitations of SCBA, as firefighters need to be aware that there are situations that may be encountered where SCBA will not provide adequate protection, and may be subject to catastrophic failure. For example, incidents involving hazardous materials other than those commonly encountered at structure fires (see USFA Major Fires Investigation Report *Sodium Explosion Critically Burns Firefighters Newton, MA.*)

In addition to initial training and certification in SCBA use, continued practice and advanced training should also be addressed to ensure that firefighters maintain proficiency. Regular drills will improve firefighters' familiarity and increase their comfort level with SCBA in hazardous situations. Some fire departments and State fire training agencies offer specialized "Smoke Diver" or "Breathing Equipment Specialist" courses to further develop firefighters' SCBA skills. These are often conducted using a train-the-trainer format, with the goal being to train instructors to teach advanced SCBA skills to firefighters and rescuers in their home jurisdictions.

Firefighters can be encouraged to perform a "buddy check" with their partner prior to entering a hazardous environment. Each partner can quickly check to ensure that the other has SCBA properly donned and operational. This way, problems can be identified and corrected before they hamper interior operations, or injure firefighters.

PROBLEM TRENDS

As noted earlier, many of the problems that have characterized SCBA failures in the past have been corrected with the advent of more realistic testing procedures and tougher standards. Modern, NFPA-compliant SCBA are extremely rugged and durable. The National Institute for Occupational Safety and Health (NIOSH) receives approximately 15-20 complaints each year about potential SCBA problems. When a complaint is received, NIOSH begins an investigation. Many of the complaints are determined by NIOSH to be the result of the complainant's unfamiliarity with, or their dislike of, the particular model of SCBA. About half of the investigations uncover a legitimate problem.

Most of the problems uncovered are minor in nature. These are often quickly and easily correctable with a safety bulletin from the manufacturer to users of the particular model of SCBA. NIOSH may also issue a "Respirator User Notice" to notify users of a problem. Most often, the problem is primarily a preventive maintenance or training issue rather than an actual design or construction defect. However, there are some trends that appear in the verified problems:

Regulators- - One of the most common complaints about modern SCBA from all manufacturers is a tendency for mask-mounted regulators to unexpectedly disengage from the facepiece. NIOSH has investigated incidents in which the evidence suggests that the reported disengagements were primarily due to operator error, as opposed to a design defect. This type of regulator will most often have a positive-locking mechanism, but the mechanism will not fully engage unless the user properly inserts the regulator into the mask and engages the mechanism. Training and familiarity with the specific type of mask-mounted regulator in use on a particular model of SCBA should minimize this problem. It is important that users of SCBA with mask-mounted regulators ensure that they are properly engaged before entering a hazardous environment, which can be easily accomplished during a "buddy check" by the other member of the entry team. The potential for regulators to freeze during cold weather operations is also real, and firefighters should be aware of this possibility.

Low-air Alarms- - Another common complaint received by NIOSH is the tendency for "whistle" –type low-air pressure alarms to blend with background noise on the fireground. The concern is that firefighters experience difficulty realizing that their alarm is sounding due to the noisy environment. Some manufacturers have handled this concern by mounting lights on the SCBA that activate with the low-pressure alarm, providing a visual as well as an audible warning to the firefighter and other members of the entry team. As mentioned previously, the 1997 edition of NFPA 1981 requires that a minimum of two low-air alarms be provided on the SCBA, addressing different senses. This enhancement should help mitigate any problems related to low-air alarm recognition and interpretation. Still, this is primarily a training issue. Firefighters should continuously monitor their pressure gauges while they are in a hazardous environment. Crucial decisions on how to conduct interior operations can only be made with knowledge of the air status of each member of the entry team. Continually advancing into a structure until the low-pressure alarm sounds is an invitation to disaster. Firefighters should be aware of their rate of air consumption while performing strenuous fireground tasks, and must ensure that they always have enough air remaining to safely exit the hazardous environment if the situation deteriorates.

Facepieces- - In the past, NIOSH has received complaints about facepiece lenses becoming opaque or melting as a result of extreme exposure to radiant heat, particularly during live-burn training evolutions. During these evolutions, firefighters and fire instructors should be especially cautious not to stress the SCBA and its components past the point where failure becomes a possibility.

The majority of the complaints received by NIOSH and by individual SCBA manufacturers can be correlated with preventive maintenance and training issues. NFPA 1981-compliant SCBA that are properly maintained and used by well-trained firefighters and rescuers should provide protection from those hazards commonly encountered during emergency situations. Additionally, it is important that fire departments continue to upgrade their breathing apparatus to meet the most current edition of NFPA 1981, to prevent repeating past tragedies.

OTHER ISSUES

Pushing the Edge- - Despite the myriad tests performed on SCBA during the certification process, there are limitations to the amount of punishment that SCBA can survive without experiencing failure or malfunction. Some of those involved in the design and construction of SCBA fear that, despite the more realistic testing procedures adopted by the NFPA, firefighters are pushing SCBA to their maximum safe operating limits and beyond. Care must be taken to ensure that SCBA are not routinely pushed past the "edge of the envelope" in which they were designed to operate. Subjected to this type of punishment on a regular basis, even properly maintained SCBA can be expected to fail. Firefighters should also be aware that situations exist where SCBA will not provide adequate protection from the environment, particularly those involving certain hazardous materials.

Evaluating the Fire Environment- - There is some concern among members of the fire service community that as firefighters are increasingly well protected, they are less able to discern changes in the environment around them, especially with regard to changes in thermal balance, and impending flashover conditions. This concern has led to a fear that firefighters are venturing into conditions where their standard personal protective equipment cannot protect them. As a result, some research and development efforts are seeking to provide affordable adjuncts that could replace firefighters' lost senses, and warn them to retreat as conditions become untenable.

Cold Weather- - NIOSH has received complaints about SCBA performance in cold weather. When such a complaint is received, NIOSH attempts to duplicate the exact conditions in which the reported problem occurred. With one exception, the results of these investigations were inconclusive. NIOSH did determine that although the regulator in one case met approval requirements, it's design was susceptible to freezing in the open position. In this case, the manufacturer modified the design and took measures to ensure that existing SCBA units were retrofitted with the newly designed regulator. It should be noted that, although SCBA are expected and approved to perform in cold weather, they do have a lower temperature limit (usually -20 to -25 degrees Fahrenheit). When temperatures approach these extremes, the moisture content of breathing air in the cylinders becomes an important issue[9], along with the handling that SCBA receive on the fireground. Ensuring that SCBA are not left sitting out in the open during periods of extreme temperatures should help prevent cold-related problems from occurring.

Improper Modification or Use- - It is important to note that this report addresses self-contained breathing apparatus intended, approved, and certified for use in structural firefighting. During the approval and certification process, SCBA are tested as complete units (including facepiece nosecups and waist straps), in the manner in which they are expected to be worn (i.e., on the firefighters back, with straps properly applied and tightened, and with all buckles buckled). Modifications or improper wearing of SCBA tested in this manner may void NIOSH approval and NFPA certification.

---

[9]Although OSHA 29 CFR 1910.134 requires the use of Grade D breathing air with a -50 degree Fahrenheit dewpoint in SCBA, some manufacturers are more stringent and recommend the use of Grade D breathing air with a lower moisture content (on the order of -65 degrees Fahrenheit). Fire departments, especially those in climates where extremely cold temperatures are expected, should use whichever requirement is more stringent, whether OSHA's or the SCBA manufacturer's.

Modifications to SCBA that may void NIOSH approval and result in less than peak performance of the unit include: the removal of nosecups (which may promote facepiece fogging in cold temperatures), cutting waiststraps, and replacing factory-installed buckles with parts from a different manufacturer. A regularly performed and effective preventive maintenance program can help identify any improper modifications to SCBA.

Improper or incomplete use of SCBA is another potential problem area. A common example of this occurs when firefighters use SCBA without having the waiststrap properly fastened and tightened. This practice voids NIOSH approval, may make the SCBA more difficult to wear, since the waiststrap is an integral part of the weight-distribution system, and can make the firefighter more prone to entanglement. Although "reduced-profile" maneuvers where SCBA may be temporarily removed from the firefighters back in an emergency situation are commonly taught, technically they too void NIOSH approval of the SCBA. In a more extreme example, using structural firefighting SCBA in confined space operations where the SCBA is not worn on the entrant's back (although it may be pulled behind, pushed ahead, or suspended above the rescuer) voids NIOSH approval and can have severe negative consequences (NIOSH has investigated deaths where SCBA were used in this manner). Finally, although it seems obvious, structural firefighting SCBA are not intended for use in an underwater environment where SCUBA (Self-Contained Underwater Breathing Apparatus) gear is the appropriate choice. Although SCBA may remain functional during an accidental, temporary submersion, an eventuality that fire department SCBA training programs should address, they are neither designed nor approved by NIOSH for prolonged use in the aquatic arena.

NEW CAPABILITIES FOR SCBA

Properly performed preventive maintenance and thorough, comprehensive user training are a key to preventing SCBA failures. Advancing technology holds promise to further reduce the potential for failures under extreme conditions faced by firefighters and rescue personnel. Although technology cannot replace proper training and preventive maintenance, the incorporation of several recent technological advances in SCBA design may help give firefighters increased protection from situations commonly encounter in structure fires. Some examples of this technology are the integration of heat-sensing devices, Heads-Up Displays (HUDs), communications interfaces, and thermal imaging devices.

In order to develop an understanding about the direction of SCBA technology, general information on SCBA trends and developments was gathered from members of the fire service and from manufacturer's representatives. A discussion of the trends identified during this research is presented here.

Factors to Consider When Adding Capabilities

Future expectations for SCBA include the widespread incorporation of heads-up displays, thermal imagers, heat sensors, video/data links, and improved communications devices. Some of these items are available now, although the expense of such equipment places it out of reach for many fire departments.

While added capabilities such as the above are generally regarded as improvements, several critical issues must be considered when adding capabilities. The first is the additional weight that the device brings to the SCBA. Many of the recent improvements in SCBA construction were directed toward decreasing the weight of the unit (e.g., carbon-fiber cylinders), to reduce fatigue, strain, and increase maneuverability. For firefighters, every ounce of added weight is a significant factor during tactical operations. The value added to the SCBA by enhanced capabilities must be balanced against the need to keep SCBA as lightweight as possible. This is a critical safety issue as the weight of the SCBA and other elements of the protective ensemble contribute heavily to fatigue.

Another important issue is the impact of added capabilities on SCBA profile or bulkiness. Profile and weight reduction efforts have gone hand-in-hand, culminating with the recent introduction of carbon-fiber breathing-air cylinders that hold more air in smaller cylinders. Adding devices to SCBA that provide additional capabilities, but which enlarge the equipment profile, may increase the likelihood that firefighters will become entangled or trapped. Streamlined, low-profile SCBA help firefighters maintain balance by minimizing changes to their center-of-gravity. This performance characteristic is especially important during roof operations, while ascending or descending ladders, and during interior firefighting or search and rescue efforts.

Successful interior firefighting operations demand that firefighters direct their complete and total attention to the task at hand; a moment of indecision or inattentiveness can spell disaster. Safely performing tasks like fire attack, ventilation, or search and rescue requires that firefighters continually monitor the conditions around them, looking for subtle clues that may signal a dramatic change like flashover or impending collapse. This process of mental size-up is hindered by such things as noise, lack of visibility, and the overall nature of the hostile fire environment. As a result, firefighters can quickly become challenged by the sheet number of sensory inputs and stimuli that they are receiving. A variety of technological add-ons to SCBA have been proposed to help firefighters monitor the environment around them including: built-in thermal sensors, heat alarms, and heads-up displays, as well as communications and data links to relay information on the "big picture" to and from the incident commander.

While these added capabilities may positively affect firefighter communication and safety, they should be carefully considered not only from a physical standpoint (e.g., weight and profile) but from a mental standpoint, as well. While technological adjuncts may provide a wealth of useful information, they also cause additional distractions for already taxed firefighters. The beeping, blinking, wailing, and whistling of alarms inside an already chaotic environment places additional stress on firefighters, and may hinder their focus on assigned tasks, causing them to miss subtle and critical clues on changing fire conditions.

Added capabilities demand additional training. With respect to SCBA, the importance of thorough, comprehensive user training cannot be overstated. Every device added to SCBA requires that firefighters be trained in its proper use, operation, and maintenance. Failure to perform such training will limit the usefulness of the device.

The unpredictable fire environment, coupled with the routinely harsh treatment received by fire equipment daily, make it difficult to develop devices that are affordable, durable, reliable, and that provide consistent service. While regular preventive maintenance programs can help keep these devices operating properly, man "cutting-edge" devices are currently unsuitable for fire service use. Problems with add-ons can also result in more down-time for SCBA units, which poses a problem to departments with a limited number of available SCBA. To avoid having to send the entire unit for repair when there are problems, firefighters may disregard or not use the device, with a negative impact on safety.

Cost is another major factor when considering adding capabilities to SCBA. While it may be possible to design and construct a durable and reliable technological adjunct, such developments may not be economically feasible for manufacturers unless many fire departments demand it on their SCBA. In turn, fire departments must also balance the value of added capabilities against the many competing demands on department resources.

Simplicity and Reliability

Improvements in the simplicity and reliability of SCBA may also be a way to add value and enhance firefighter safety. By making SCBA simpler and easier to use and maintain, failures related to inadequate training or preventive maintenance should be further reduced. Simpler, more reliable SCBA should also help fire departments conserve their fiscal resources by increasing service life and reducing maintenance and repair costs.

Several manufacturers contacted indicated that their major emphasis during the next decade will be on improving the simplicity and reliability of SCBA, as opposed to incorporating technological adjuncts. Particular attention will probably be directed toward using ergonomic design principles to enhance the "wearability" of SCBA.

Potential Areas of Improvement

Some of the improvements described here have already been incorporated in SCBA, and are currently available to fire departments wishing to upgrade or replace existing units. A variety of other technological advances are being studied or tested. Ultimately, it is fire departments who determine what improvements the manufacturers incorporate into SCBA, and at what price.

Masks and Facepieces- - Recent trends in ergonomic design have led to a variety of recent advances in the design and construction SCBA facepieces. Improvements have been made to enhance the clarity of voice communications, reduce fogging, improve peripheral vision, and provide a range of sizes to ensure that firefighters are properly fitted. Still, the basic shape and construction of SCBA facepieces has remained relatively unchanged in recent years.

Polycarbonate remains the material of choice for SCBA facepieces since it offers the best combination of thermal protection and clarity currently available. Some recent experiences with radiant heat transmission through the lens of polycarbonate facepieces may indicate that further research is needed to develop an alternative material that better resists the high radiant heat release from fires.

Some manufacturers are studying the benefits of integrating helmets and facepieces, although there is some question about whether or not an integrated helmet/mask system meets current NIOSH regulations and NFPA standards; this is due to concerns about the integrity of the face-to-mask seal if the helmet were struck by an object hard enough to knock it off the wearer's head, thus potentially exposing him/her to a hazardous atmosphere. Such integrated helmets have long been used by the French Fire Service, and are the standard helmet for most French departments.

Heads-Up Displays- - Although not yet widely available, another advance being studied is a Heads-Up Display (HUD) for SCBA. This feature continuously provides the user with critical information on their life support status, without the need to divert attention to a remote pressure gauge or guess about whether or not their facepiece is leaking significantly enough to cause a problem. HUDs may also be developed to transmit information from the command post to firefighters operating in remote locations. At least one manufacturer currently offers a simplified version of a HUD using a series of lights on the facepiece to indicate remaining air pressure and positive-pressure status.

The difficulties and expense of engineering a full-featured display as part of an SCBA facepiece will probably preclude HUDs from being incorporated in facepieces anytime in the near future, although they may be marketed as add-ons that attach to the helmet. Such a device is already available from one company (not an SCBA manufacturer) although it is not intended for use in interior firefighting. The use of such a device, even if not technically part of the SCBA itself, will have an impact on the SCBA, especially with regard to firefighters' field-of-vision and their ability to process SCBA-related information. HUDs are widely used in aerospace and military applications, and it seems likely that some form of these devices will be incorporated into firefighters' personal protective ensemble in the future.

Thermal Imaging- - Although the value of being able to "see through smoke" is clear to firefighters, the cost and complexity of currently available thermal imagers make their widespread incorporation into SCBA facepieces rather unlikely, at least in the near future. As in the case of HUDs, thermal imaging devices will probably continue to be marketed as adjuncts that attach to the helmet. Also like HUDs, the thermal imaging system, while not technically part of the SCBA, may directly affect its performance by changing the firefighter's field-of-vision, or by adding additional weight to the helmet of SCBA harness.

Communications- - Several manufacturers offer facepieces with integrated communications systems. These can be of different types including: voice amplifiers, two-way radio interfaces, wireless intercoms, or a combination. The benefit of integrated communications is easy to understand as fireground communication has long suffered from the difficulty of attempting to speak and be understood through an SCBA facepiece, often in a loud environment. The difficulty of adapting microphones and other hardware to work inside the SCBA facepiece has made voice clarity hard to achieve. The cost of these systems has also reduced their appeal, as few fire departments can afford to outfit all of their SCBA. While various types of integrated communications systems have been available for several years, they may become more commonplace as their durability and reliability continue to improve.

Breathing Air Cylinders- - Carbon-fiber SCBA cylinders are currently available from several manufacturers and represent a major advancement in the "state-of-the-art". These cylinders have enabled weight and profile reductions without having to sacrifice the volume of air contained within the cylinder. As a result, firefighters can carry more air, with less weight on their backs, and less chance of becoming entrapped or entangled. Although breathing air cylinders with durations exceeding one hour are now possible, fire departments should carefully evaluate whether or not it is prudent to use them in the structural firefighting environment.

While carbon-fiber cylinders may feel very different from ordinary composite cylinders in terms of weight, they are constructed in essentially the same manner, and are still susceptible to physical and chemical damage or abuse.

Several manufacturers are studying ways to further reduce the profile of SCBA by using differently shaped containers or flexible bladders within lightweight cylinders. The problems of gaining DOT certification for these designs indicate that, currently, carbon-fiber represents the best available technology for SCBA cylinders.

Monitoring and Diagnostics- - An area where added capabilities are likely in the near term is in monitoring and diagnostics. The widespread incorporation of integrated PASS devices is one example of how monitoring technology is currently being integrated in SCBA. It is likely that the integrated PASS will evolve into a "black box" that performs several different functions. Future advances will probably include the use of technology to monitor a variety of SCBA performance factors and provide an interface through which maintenance personnel can download diagnostic information on the unit. Advanced diagnostics will allow preventing maintenance to be performed based on the actual use of the unit in terms of hours, instead of by guidelines applied to a set of SCBA without regard for how often or how hard, they are used.

Sensors may also be incorporated into SCBA to measure outside temperature, and sound an alarm if the user enters an environment where their protective equipment may be inadequate. Firefighter location systems using radio or GPS (Global Positioning System) transceivers may also become commonplace in future equipment configurations.

The monitoring and diagnostic functions incorporated into future SCBA may also be able to monitor human performance factors like the number of breaths per minute, heart rate, skin temperature, etc., helping to identify problems before they can contribute to firefighter casualties.

The areas of improvement described here are general in scope and SCBA manufacturers continue to develop technology to enhance future SCBA. Additional research is being performed by the fire service in cooperation with major universities and other agencies like NASA and the Department of Defense. Although it is important to remember that even the most fail-safe, state-of-the-art SCBA still depends on firefighters for its proper operation and maintenance, technological improvements hold great promise for helping to prevent SCBA failures.

RECOMMENDATIONS

In summary, the following actions are highly recommended to prevent SCBA failures leading to firefighter injuries.

Preventive Maintenance- -Regular preventive maintenance will help ensure the safety and well-being of firefighters and compliance with applicable regulations. Fire departments should implement SCBA preventive maintenance programs. Regular maintenance by qualified personnel will help prevent SCBA failures and should be conducted using manufacturers recommendations as a minimum standard. **SCBA subjected to extreme levels of use should be maintained and tested at more frequent intervals.**

User Training- - Firefighters must receive comprehensive training in the use and limitations of SCBA before being exposed to hazardous environments. In addition to initial training, continued drill and practice should be conducted to maintain proficiency with SCBA. Training should also be provided on proper field maintenance and cleaning procedures. Thorough, ongoing user training should help decrease the incidence of low-order SCBA failures. **Training should emphasize the limitations of SCBA to help ensure that firefighters do not put themselves in situations where SCBA will not provide adequate protection from the hazardous environment.**

Problem Recognition- - Many of the problems reported to NIOSH annually are related to preventive maintenance or training deficiencies. Improper use or modification of structural firefighting SCBA may have negative consequences. Firefighters and maintenance personnel should be able to recognize potential problems that may lead to failures of SCBA or breathing air cylinders. **Any SCBA or breathing air cylinder that has been dropped, run-over, burned, submerged, exposed to chemicals, or otherwise mistreated should immediately be purged from the service for examination and testing by qualified repair personnel.**

Regular Upgrades- - Fire departments should upgrade or replace SCBA to meet the current edition of the NFPA 1981 *Standard on Open-Circuit Self-Contained Breathing Apparatus for Fire Fighters.* **SCBA with demand-type regulators do not provide firefighters with adequate respiratory protection and should be replaced with positive-pressure type units.**

Use of Standards- - The NFPA standards summarized previously in this report can be very useful to fire departments developing SCBA programs. Programs meeting the requirements contained in these standards are likely to be compliant with applicable Federal or State regulations.

Pursuit of Technology- - Fire departments should continue to encourage the development and incorporation of new technology in SCBA design and construction. The use of new materials and construction methods holds the promise of stronger, lighter, more capable SCBA providing firefighters with increased safety and protection from hazardous environments.

Record-keeping- - The importance of keeping accurate, up-to-date records of preventive maintenance actions, individual SCBA and cylinder test results, and SCBA user training cannot be overstated. Air quality records should also be maintained to ensure that firefighters receive clean, dry air in their SCBA cylinders. Accurate records can be used to show trends that may justify performing regular maintenance at intervals more frequent than those recommended by the manufacturer.

APPENDIX A

"Acidic Fluid Leads to Air Cylinder Failure", Fire Engineering, November, 1996

Reprinted with permission from the editors of Fire Engineering.

Appendix A (continued)

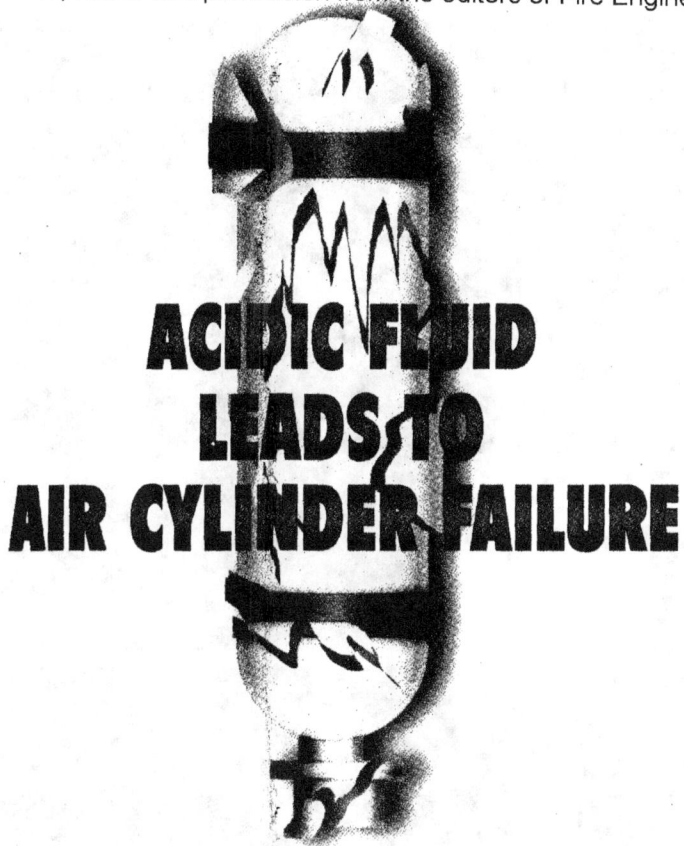

At approximately 9:14 p.m., on March 4, 1996, a DOT-E- 8059, 4,500-psi, fiberglass-wrapped composite aluminum breathing air cylinder ruptured without warning while stored in an equipment compartment on a reserve engine (Engine 1-7) of the Humboldt No. 1 Fire Protection District in Eureka, California. Engine 1-7 suffered approximately $5,000 worth of damage; and Engine 1-4, adjacent to Engine 1-7, sustained $600 in damages, predominantly minor dents and scratches in the paint. Engine 1-7, a 1,000-gpm pumper with equipment compartments, hosebed, and crew cab with jump seats in conventional arrangement, housed the cylinders in two left-side rear compartments, aft of the pump panel. Fortunately, no one was injured in the incident.

Air cylinders were found six and 10 feet away, and there were indications that they had been airborne that distance. The compartment doors were severely damaged; the rear door was completely separated from the engine body and lay on the floor, curled into a "U" shape. The overpressure caused the diamond-plate top of the compartment to expand up and outward, twisting it and pulling it off the bolts holding it in place. Also, the rear seam of the compartment floor spread open, leaving a one-half-inch gap between the compartment floor and the back wall. Directly behind the location of the ruptured bottle, on the compartment's back wall, was a dent about six inches in diameter and one inch deep, resulting from a direct hit from the bottle.

The force of the rupture caused the cylinder—a half-hour-capacity fully-wrapped composite cylinder (manufactured by EFI Corporation) and part of an SCBA manufactured by International Safety Instruments (ISI)—to split open and its walls to violently shred the plywood backing board in the compartment. The ruptured cylinder was found on the apparatus room floor, in front of its holding bracket and almost directly beneath the damaged cylinder tube door on the other engine. Several of the cylinder valves showed serious damage after the incident and were immediately determined to be unfit for service.

At the time of the rupture, the on-duty firefighters were in the station day room. The rupture shook the station and created a few moments of concern. Captain Dave Gibbs investigated and ordered firefighters out of the apparatus room. After 30 minutes and following a telephone conference with Chief Robert Heald, Gibbs donned full turnouts, a helmet, and a flack jacket and entered the apparatus room. He opened the cylinder valves on the involved cylinders to bleed off the pressure. He left the area and waited for the cylinders to drain their contents before allowing his crew to enter the area. Meanwhile Gibbs radioed to dispatch that his station was out of service.

After another 30 minutes, Gibbs and his crew entered the apparatus room to survey the damage.

INVESTIGATION

The sales manager and chief engineer of EFI Corporation, representatives of CAL-OSHA and the Department of Transportation (DOT), marketing and sales managers from ISI, safety experts, insurance adjusters and inspectors, breathing apparatus users from several disciplines including the U.S. Coast Guard and the U.S. Department of Energy, representatives of large and small fire departments, and the news media all visited the site beginning the day after the incident.

ISI notified EFI Corporation of the cylinder failure the same evening of the incident. Since the cause of the cylinder's failure was unknown at the time, EFI Corporation, issued a quarantine for the 199 other cylinders manufactured in the same lot. The quarantine was rescinded when the investigation subsequently revealed that the cylinder rupture was the result of the cylinder's coming in contact with a highly acidic cleaning fluid.

EFI Corporation contracted with Failure Analysis Associates, Inc. (FaAA) to investigate the incident. The failed cylinder and the cylinder storage rack holding the cylinder at the time it failed were transported to FaAA, which undertook an investigation that included information gathering, visual examination, optical and scanning electron microscopy, chemical analysis, stress corrosion cracking testing, and chemical exposure testing.

The investigation revealed that the Humboldt Fire Department SCBA cylinders had been used for a training exercise. The following sequence of events was reconstructed:

• On February 27, the SCBA cylinders were used during a multiagency fire training

Appendix A (continued)

(Top left, top right, bottom left) The failed cylinder caused significant damage to the engine on which it was stored and minor damage to the adjacent engine. The damage caused by the failed cylinder (on the floor between the two parked engines) included blowing off the door of the compartment in which the cylinder was stored, blowing open and deforming an adjacent storage compartment door, expelling the majority of the contents of the compartment, and dismounting and throwing to the floor 10 feet of large hard-suction hose. (Bottom right) Close-up of damaged cylinder. The cause of the cylinder's failure was a previous contact with a highly acidic fluid that chemically attacks fiberglass. (Photos by David Gibbs.)

exercise for the officers of Humboldt County Sheriff's Office and several correctional facilities. Nine fire personnel from Humboldt No. 1 Fire Protection District served as technical advisors and safety support personnel.

• On the evening of February 27, a deputy sheriff with the Humboldt County Sheriff's Department transported the cylinders in his flatbed trailer to the Eureka Fire

Department for refilling. The cylinders were in the trailer about 30 minutes. One fiberglass cylinder and four to five metal cylinders were on the floor of the trailer; other cylinders were on top of them.

• During loading or transport, a 12-ounce spray bottle on the floor of the trailer, containing about four ounces of aluminum cleaning fluid, broke and spilled its contents.

• After delivering the SCBA cylinders to

the Eureka Fire Department, the deputy emptied and washed his trailer.

• About two hours later, the deputy retrieved the refilled bottles from the Eureka Fire Department and returned them to the Humboldt Fire Department.

The potential exposure of a fiberglass SCBA cylinder to a corrosive fluid was a key piece of information in the failure investigation.

Appendix A (continued)

FaAA reported the following after its investigation:

• The macroscopic and microscopic fracture surface is characteristic of stress corrosion cracking in fiberglass composites.

• The results of the chemical analysis indicate that fiberglass samples of the failed cylinder contained 2-butoxyethanol, a chemical constituent of a highly acidic fluid that chemically attacks glass. The fluid is a commercially available aluminum cleaner used also for chrome and other metal surfaces that contains among other ingredients hydrofluoric acid, phosphoric acid, sulfuric acid, ethylene glycol, monobutyl ether, and nonylphenoxypoly (ethyleneoxy) ethanol.

• The rapid failure of specimens in stress corrosion cracking experiments provides qualitative verification that the fluid caused stress corrosion cracking of the accident cylinder. The appearance of fracture surfaces on the accident cylinder was very similar to the stress corrosion cracking observed on the sample exposed to the acidic fluid.

• Personal interviews indicated that during the training exercise on February 27, a fully wrapped composite cylinder was exposed to an aluminum cleaning fluid, which was likely the one in the container on the trailer. The cylinder ruptured about six days after exposure to the fluid.

The FaAA report noted: The cylinder failure "... is consistent with environmental stress-assisted cracking of the fiberglass composite (overwrap). Glass fibers failed due to the combined action of an acidic chemical environment and the stress caused by the internal pressure. This combination of chemical attack and stress acting over several days led to the failure of the cylinder."

NIOSH ADVISORY

The National Institute for Occupational Safety and Health (NIOSH) has issued an advisory that includes the following precautions:

• End-users of all composite fiberglass-wrapped cylinders could experience similar cylinder failure if they fail to take precautions to ensure that acidic cleaning fluids or other acidic chemicals do not come in contact with such cylinders. Besides contact with acidic cleaning agents in common use, NIOSH noted, fiberglass-wrapped cylinders could also be exposed to acidic chemicals in chemical plants, warehouses, and other hazmat sites.

• Fiberglass-wrapped composite cylinders should only be used, handled, cleaned, maintained, and transported by individuals who have been made aware of the need to avoid exposing such cylinders to acidic materials.

• If a fiberglass-wrapped cylinder is suspected to have come in contact with an acidic chemical, the cylinder should be depressurized immediately and removed from service. The SCBA manufacturer should then be contacted for further instructions.

Additional information is available from Robert S. Frankle, managing engineer, Failure Analysis Associates, 149 Commonwealth Drive, Menlo Park, CA 94025. ■

References

Report, Chief Robert Heald, Humboldt No. 1 Fire Protection District, Eureka, California.

"Investigation of the Failure of an SCBA Cylinder," Robert S. Frankle, Harry F. Wachob, Ph.D.; Failure Analysis Associates, Inc., Menlo Park, California, July 2, 1996.

Letter, David C. Woodward, Vice President & General Manager, EFI Corporation, Fremont, California, July 19, 1996; EFI Corporation, "Incident Report Conclusion," March 21, 1996.

"NIOSH Respirator Users Notice," National Institute for Occupational Safety and Health, Morgantown, West Virginia, July 30, 1996.

APPENDIX B

Cavallari, Bruce J., "Hydrostatic Testing of Compressed Gas Cylinders", *Fire Engineering*, October 1996

Reprinted with permission from Mr. Bruce J. Cavallari and *Fire Engineering*.

Appendix B (continued)

Reprinted with permission from the author and Fire Engineering

HYDROSTATIC TESTING OF COMPRESSED GAS CYLINDERS

BY BRUCE J. CAVALLARI

Can an SCBA cylinder be hydrotested prior to the expiration of the last hydro date? Should fiber-reinforced composite cylinders be filled in water? Do hoop or fully wrapped SCBA cylinders have a life span? This is a sampling of the questions asked every day throughout the fire service concerning compressed gas cylinder maintenance. The answers may not depend solely on region, training, and available (reliable) information. There are finite rules and regulations concerning cylinder safety, and it is *our* responsibility to be familiar and in compliance with these standards. They were created for our safety.

....

■ **BRUCE J. CAVALLARI** has served in career and volunteer organizations during his 18-year fire service career and is a member of the Palm Beach County (FL) Fire-Rescue, where he is a lieutenant in the operations division. He also heads the consulting and instruction company Environmental Safety Specialists, specializing in the use of compressed gas equipment (SCBA, liquid O_2, gas manifolds), and is a consultant for American Hydro-Test, Inc., a cylinder requalification facility in Fort Lauderdale, Florida. Cavallari is a Florida state-certified company officer, a company officer instructor, a fire safety inspector, an inspector instructor, and a minimum standards instructor.

WHAT IS HYDROSTATIC TESTING?

Hydrostatic testing is the standard method of testing cylinders in the compressed gas industry. The water jacket method of testing cylinders consists essentially of enclosing the cylinder, suspended in a jacket vessel, and measuring the volume of water forced from the jacket on application of pressure to the interior of the cylinder. The pressure applied is five-thirds the operating pressure unless otherwise specified by the manufacturer. Therefore, a 4,500-pound-per-square-inch (psi) cylinder would be tested at 7,500 pounds psi. This method is used to determine the elastic expansion, which is directly related to the cylinder's average wall thickness.

Department of Transportation (DOT) regulations require that each cylinder be inspected periodically internally and externally, calibrated testing equipment and test gauges be used, and every cylinder used for the storage of compressed gases such as air or oxygen be retested. The exact laws are in the *Code of Federal Regulations* (CFR), Part 49, subpart 173.34(e). The DOT requires each testing facility to submit to regular inspections; and each facility must maintain records of all tests performed, all persons performing the tests, and regular training records of personnel.

The DOT was established in 1967 and assumed the responsibility for the safety regulations formerly administered by the Interstate Commerce Commission.

CYLINDER DESCRIPTIONS

I'm sure we all can recognize a cylinder when we see one, but exactly how many different types of cylinders are we dealing with in our stations? Fire extinguishers are cylinders. Oxygen used on rescue vehicles is in cylinders. SCBAs and SCUBAs have cylinders. Large storage cylinders are on the air trailer/compressor. The cutting torch has cylinders. Air supply for the vehicles' air horn may come from a stored pressure air cylinder. All of these and more, depending on your specific locale, are cylinders that may require hydrostatic testing at a DOT-authorized cylinder requalification facility. Consider each type carefully.

Oxygen

Oxygen systems for EMS, forcible entry, and extrication uses can be found onboard the engines, rescues, ambulances, and staff cars throughout your workplace. Within the stations, oxygen-transfilling systems are commonplace. Whatever the cylinder size or use, hydrotesting is required every five years. Even the oxygen cylinder on the cutting torch must be retested.

Air

Air cylinders are used as our SCBA bottles, SCUBA tanks, and air-horn and air-chisel supply. All composite cylinders are retested every three years, and all aluminum and steel cylinders are retested every five years.

SCBA cylinders commonly are found in 20-, 40-, 45-, and 90-cubic-foot aluminum, steel, or composite construction. Seven- and 17.5-cubic-foot packs are also available.

Extinguishers

Even fire extinguishers can need hydrotesting. Stored pressure extinguishers such as dry chemical ABCs need retesting every 12 years. CO_2 cylinders and pressurized water extinguishers are retested every five years. The original manufacture date often is stamped into the cylinder or listed on the identification label.

TEST FACILITY RESPONSIBILITY

Cylinder requalification facilities must follow the DOT guidelines as specifically as described in the *Code of Federal Regulations* (CFR) for the cylinders being tested. They must properly inspect, record, and test each cylinder and stamp the cylinder neck or affix along the barrel a sticker indicating the facility's permit number and the month and year in

Appendix B (continued)

which the test was completed.

DEPARTMENT RESPONSIBILITY

Low- and high-pressure compressed gas storage systems are used for life-support functions. Each should be inspected thoroughly at least weekly, if not daily. Your inspection should include the recording of the pressure and a comparison of that pressure with previous readings, visualizing the hydrostatic test date, and examining the overall condition of the cylinder's exterior. This examination should include the valve body and cylinder bottom. Measure or immediately clean deep pits or gouges. Pits or gouges indicate a serious structural problem. Remove the cylinder from service, and have it tested or repaired at your hydro facility.

PROMPT MAINTENANCE

Immediately report any indications of damage. Remove the cylinder from service, and tag it with specific information that indicates the exact problem. If fiber threads have become unglued, they can be re-epoxied by qualified cylinder maintenance personnel at a hydrotesting facility. Do not overlook the quantity and depth of the loose fiber or pits.

WARNINGS

You should be aware of the following:

• All composite cylinders—hoop or fully wrapped—have a 15-year life span from the date of original manufacture. There are absolutely no exceptions. If you own a cylinder that was originally made in January 1981, its life span would have expired on January 1, 1996. Composite cylinders were just becoming popular in 1981, so keep your eyes open. Expired cylinders should begin to surface soon.

Hydrostatic testing is designed to reflect the safety of the cylinders we use. By ensuring the integrity of the cylinder, hydrotesting allows firefighters a measure of confidence while wearing what amounts to a 2,000-pound bomb on their backs.

* * *

Question: When can cylinders be tested every 10 years instead of every five years?
Answer: When the testing facility has affixed a five-point star after the last hydrotest date, the cylinder can be retested 10 years after that date. This is applicable only to certain steel cylinders. Specific requirements must be met before a star will be applied. Consult with your supplier to consider this option.

Question: Do acetylene gas cylinders need to be hydrostatically tested?
Answer: Their construction and extremely low pressure exempt acetylene cylinders from hydrostatic testing. To be certain a cylinder needs retesting, check the DOT markings on the neck or barrel. If the cylinder had a previous test date, it will need retesting. No mark, no retest.

Question: Should composite SCBA cylinders (hooped or fully wrapped) be filled in water?
Answer: I personally and some manufac-turers say no, for three reasons:

• Water could seep in between the fiberglass wrapping and the aluminum and begin a degradation process that could degrade the cylinder's aluminum or fiber, or both.

• If refilled properly, heat buildup is minimal and not damaging.

• Using water in a cooling vat will not prevent a cylinder from rupturing; cannot contain any of the debris; and could be introduced into the cylinder via the open valve, causing a problem with internal corrosion or air contamination.

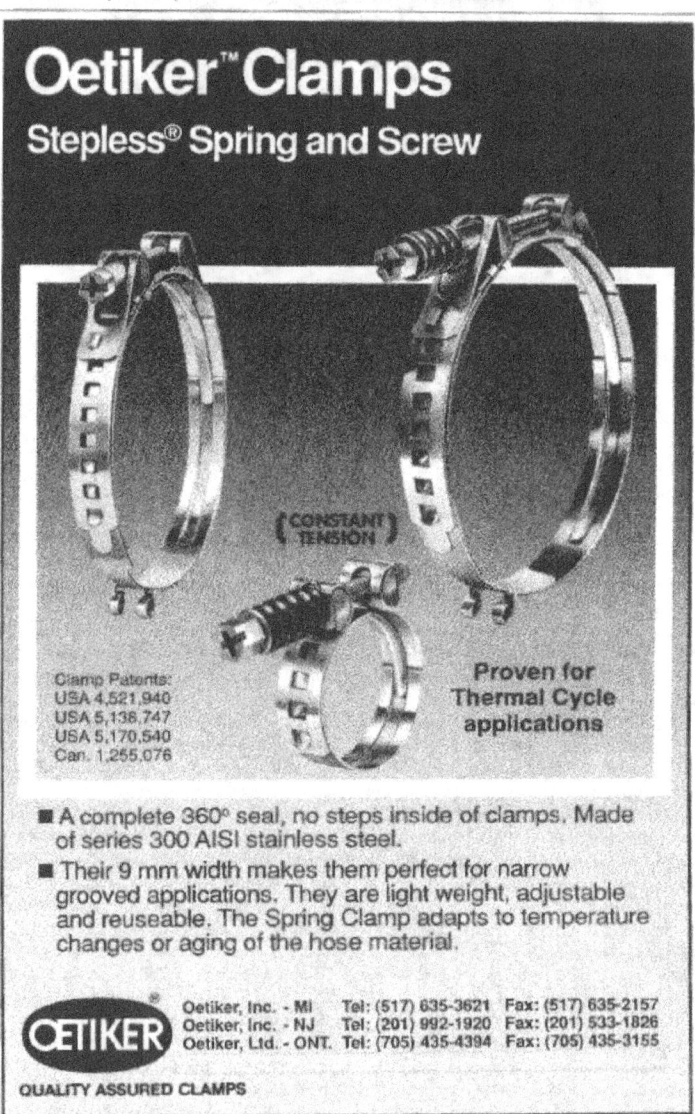

Appendix B (continued)

Question: Can a cylinder be retested more frequently than at its required interval (three, five, 10, or 12 years)?

Answer: Yes. Whenever a cylinder's integrity is in question due to impact, due to excessive heat (over 170°F), or when there is any uncertainty about the cylinder's structural makeup, the old saying "Better safe than sorry" applies.

Question: If a cylinder fails hydrotesting, should the test facility drill a hole into the cylinder, rendering it useless, or stamp into the metal the words "Failed Hydro/VIP (Visual Inspection)"?

Answer: Without the specific permission of the cylinder owner to enact such markings, the responsibility of the test facility is only to mark the cylinder in a nondamaging way and record the disposition of the cylinder in the test log. Remember, the cylinder does not belong to the test facility; no right of ownership exists. The only markings on the cylinder should be those specifically required by the DOT/CFR standard or the owner of the cylinder.

Question: What about the cylinder valve and contents pressure gauge? At the time of the hydrotest, is it required that the valve be overhauled or the contents gauge calibrated?

Answer: No. Neither the hydrotest nor any known manufacturer requires that the valve be overhauled or the gauge recalibrated at the time of retesting. Some manufacturers do "recommend" that preventive maintenance be performed on a regular basis, but the timing of this maintenance is entirely up to the end user (that's you, not the test facility). Valve overhauls and gauge recalibrations are necessary only when their accuracy is in question. If a valve overhaul is needed, be sure the facility making the repairs is authorized by the manufacturer to service the equipment.

Question: Cylinders that have exceeded the 15-year life span should be returned to the manufacturer for rewrapping, which will give them another 15 years of service life. True or False?

Answer: False. There is absolutely, positively, and without a doubt no way to refurbish a composite cylinder. To replace this cylinder, the only course of action would be to remove and save the valve from the expired cylinder. Purchase a new cylinder shell from the original manufacturer (be sure to specify that you wish to purchase only the cylinder, not the valve), and reinstall your old valve into the new cylinder. This could save you upward of $250 on the replacement of a cylinder with valve.

What can you do with your expired composite cylinders after the 15th year? Good question. They look like good cylinders and may even hold pressure just as before. Structural decomposition after 15 years of use can be substantial. The fiberglass reinforcement may be decaying, and the aluminum may be corroding. The problem is that none of this decomposition is visible beneath the fiberglass wrapping. Dispose of these cylinders any way you see fit, but be aware of the liability you can incur by reselling a condemned cylinder.

● The second problem you should be aware of is neck leaks. Aluminum cylinders may develop cracks or pinhole leaks at the neck near the valve assembly. While filling, give special attention to this area, especially since the life span issue is beginning to develop. Immediately empty cylinders with cracked necks, and return the cylinders to the manufacturer for evaluation as soon as possible. (Don't forget to remove your valve.) ■

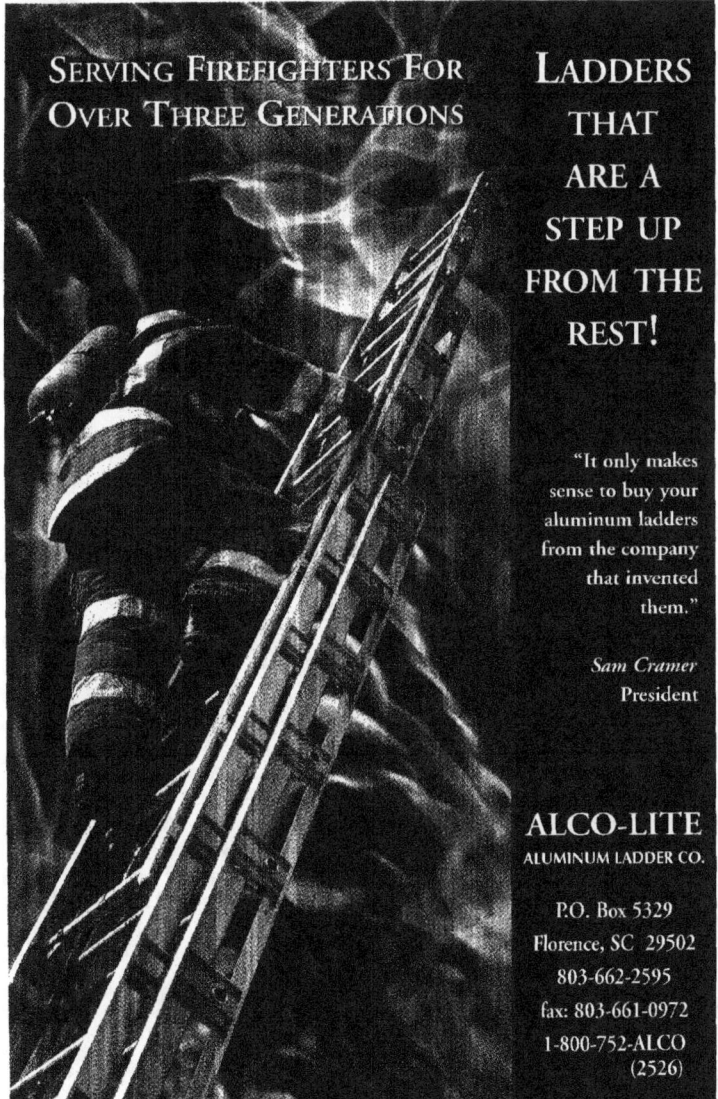